水 星

高速星球

MERCURY

The High-Speed Planet

（英国）埃伦·劳伦斯／著　刘 颖／译

江苏凤凰美术出版社

著作权合同登记图字：10-2022-144

图书在版编目（CIP）数据

水星：高速星球 /（英）埃伦·劳伦斯著；刘颖译．

南京：江苏凤凰美术出版社，2025. 1. --（环游太空）.

ISBN 978-7-5741-2027-3

Ⅰ. P185.1-49

中国国家版本馆 CIP 数据核字第 20244NY787 号

策　　划	朱　婧
责任编辑	高　静　奚　鑫
责任校对	王　璇
责任设计编辑	樊旭颖
责任监印	生　嫄
英文朗读	C.A.Scully
项目协助	邵楚楚　乔一文雯

丛 书 名	环游太空
书　　名	水星：高速星球
著　　者	（英国）埃伦·劳伦斯
译　　者	刘　颖
出版发行	江苏凤凰美术出版社（南京市湖南路 1 号　邮编：210009）
印　　刷	南京新世纪联盟印务有限公司
开　　本	710 mm × 1000 mm　1/16
总 印 张	18
版　　次	2025 年 1 月第 1 版
印　　次	2025 年 1 月第 1 次印刷
标准书号	ISBN 978-7-5741-2027-3
总 定 价	198.00 元（全 12 册）

书中加粗的词语见词汇表解释。

Words shown in **bold** in the text are explained in the glossary.

欢迎来到水星

Welcome to Mercury

想象一个距离地球1亿多千米的星球。

Imagine a world that is over 100 million kilometers from Earth.

那里的地面布满了岩石和大大的坑洞（陨石坑）。

In every direction, the land is rocky and covered with large holes called **craters**.

在那里，太阳看起来有地球上的3倍大。

The Sun looks three times larger than it does from Earth.

白天，那里温度极高，甚至能使某些金属融化。

During the day, it is so hot that some metals would melt.

晚上，那里比地球上最冷的地方冷得多。

At night, it is much colder than the coldest place on Earth.

欢迎来行星水星！

Welcome to the **planet** Mercury!

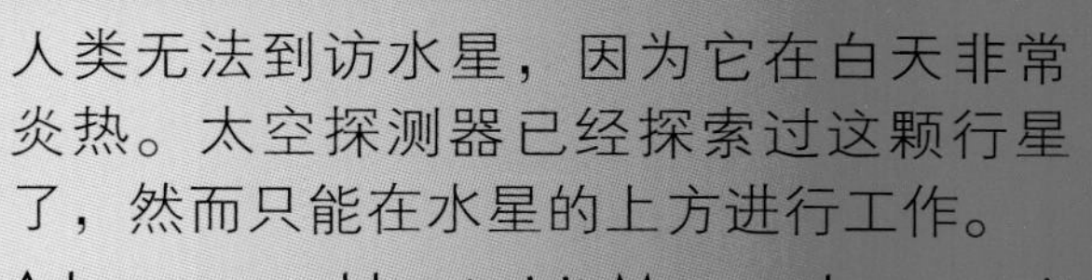

人类无法到访水星，因为它在白天非常炎热。太空探测器已经探索过这颗行星了，然而只能在水星的上方进行工作。

A human could not visit Mercury because it is too hot in the daytime. Spacecraft have explored the planet, however, from just above its surface.

这张水星表面陨石坑的照片是由一个名为“信使号”的太空探测器拍摄的。

This picture of the craters on Mercury's surface was taken by a spacecraft called *Messenger*.

太阳系 The Solar System

水星以约170 500千米每小时的速度在太空运动。

Mercury is moving through space at about 170,500 kilometers per hour.

它沿着一条椭圆形（或蛋形）的轨道绕太阳运动。

It is moving around the Sun in an oval, or egg-shaped path.

水星是围绕太阳公转的八大行星之一。

Mercury is one of eight planets circling the Sun.

八大行星分别是水星、金星、我们的母星地球、火星、木星、土星、天王星和海王星。

The planets are called Mercury, Venus, our home planet Earth, Mars, Jupiter, Saturn, Uranus, and Neptune.

冰冻的彗星和由岩石构成的小行星也围绕着太阳公转。

Icy **comets** and rocky **asteroids** are also moving around the Sun.

太阳、行星和其他天体共同组成了“太阳系”。

Together, the Sun, the planets, and other space objects are called the **solar system**.

小行星是巨大的太空岩石。太阳系中的大多数小行星都集中在被称为“小行星带”的环状带中。

Asteroids are huge space rocks. Most of the asteroids circling the Sun are in a ring called the asteroid belt.

小行星 **An asteroid**

太阳系 The Solar System

水星是距离太阳最近的行星。
Mercury is the closest planet to the Sun.

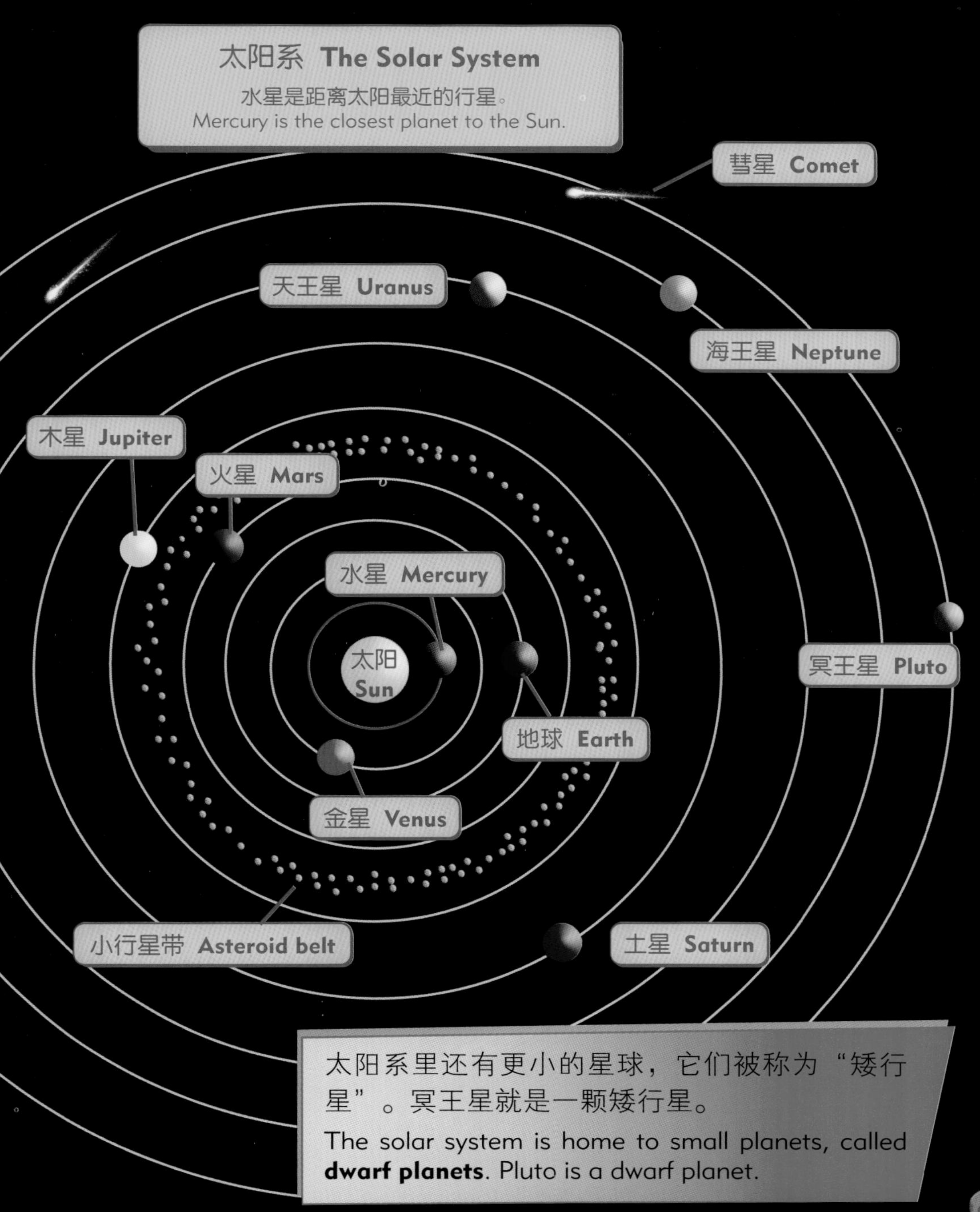

太阳系里还有更小的星球，它们被称为“矮行星”。冥王星就是一颗矮行星。

The solar system is home to small planets, called **dwarf planets**. Pluto is a dwarf planet.

水星的奇幻之旅

Mercury's Amazing Journey

行星围绕太阳公转一圈所需的时间被称为“一年”。

The time it takes a planet to **orbit**, or circle, the Sun once is called its year.

地球绕太阳公转一圈需要略多于365天，所以地球上的一年有365天。

Earth takes just over 365 days to orbit the Sun, so a year on Earth lasts 365 days.

水星比地球离太阳近得多，所以它绕太阳公转的路程更短。

Mercury is much closer to the Sun than Earth, so it makes a much shorter journey.

水星绕太阳公转一圈大约需要88个地球天。

It takes Mercury about 88 Earth days to orbit the Sun.

这意味着，地球上的1年，相当于水星上的4年！

This means that for every year on Earth, Mercury has about four years of its own!

当行星围绕太阳公转时，它也像陀螺一样自转着。

As a planet orbits the Sun, it also spins, or **rotates**, like a top.

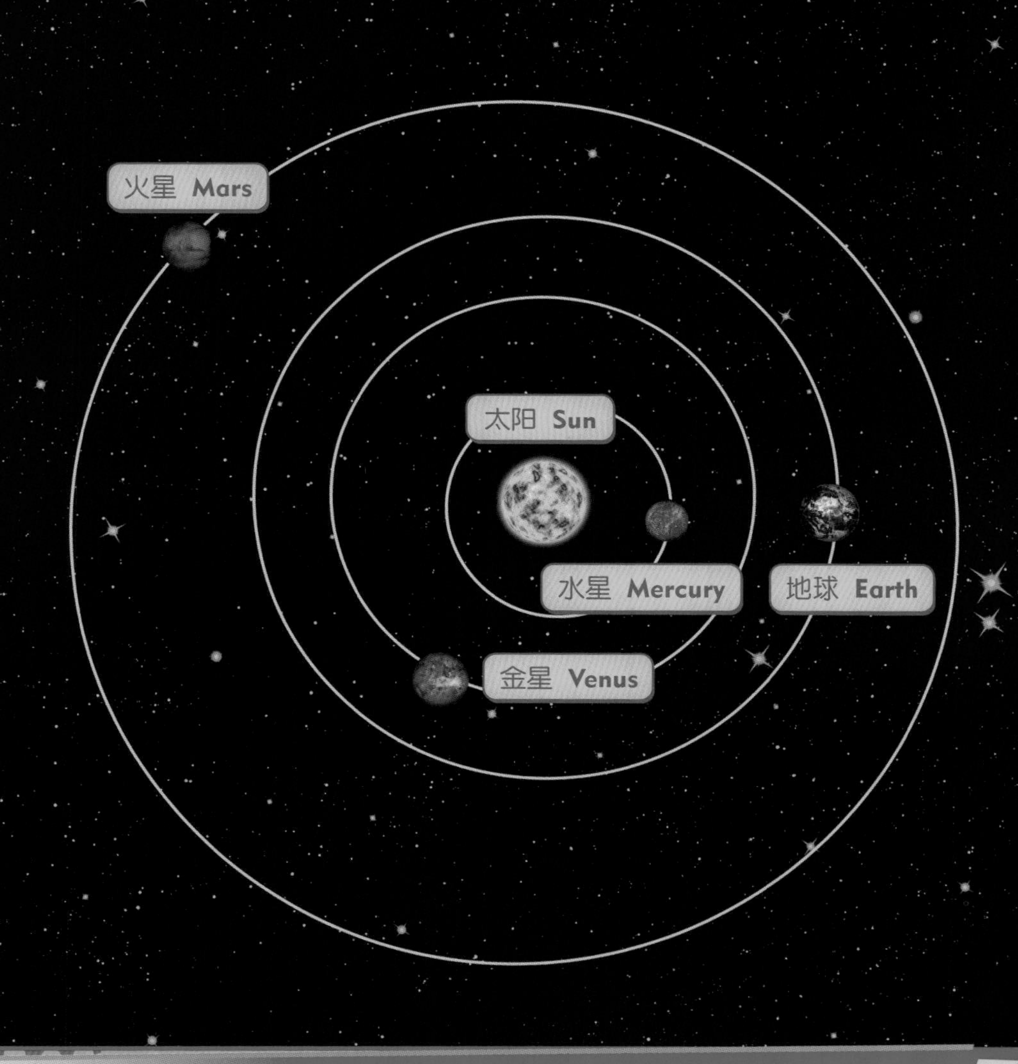

地球绕太阳公转一圈的路程约为9.4亿千米，而水星绕太阳公转一圈的距离大约仅为3.6亿千米。

To orbit the Sun once, Earth makes a journey of about 940 million kilometers. Mercury's journey is only about 360 million kilometers.

近距离观察水星

A Closer Look at Mercury

水星是太阳系中最小的行星。

Mercury is the smallest planet in the solar system.

它比地球小得多，比地球的卫星（月球）略大。

It is much smaller than Earth, and just slightly larger than Earth's Moon.

我们的地球家园覆盖着一道厚厚的气体层，称为“大气层”。

Our home planet, Earth, is covered with a thick layer of gases called an **atmosphere**.

正是这层气体使地球的天空看起来是蓝色的。

It's these gases that make Earth's sky look blue.

水星不像地球那样有厚厚的大气层。

Mercury does not have a thick atmosphere like Earth's.

所以，无论是白天还是黑夜，水星的天空看起来都是黑色的。

So from Mercury the sky looks black whether it is day or night.

行星的大气层能阻挡部分太阳热量。水星几乎没有大气层，而且它距离太阳很近。这就是水星在白天变得如此炎热的原因。

A planet's atmosphere blocks some of the Sun's heat. Mercury has almost no atmosphere, and it is close to the Sun. That's why Mercury gets so hot in the daytime.

行星的大气层能在白天吸收太阳热量，并在夜晚发挥保温作用。水星在夜晚极其寒冷，是因为所有在白天吸收的热量都散发掉了。

At night, a planet's atmosphere traps some of the daytime heat. Mercury is super cold at night because all its heat escapes.

小行星，大陨石坑

Small Planet, Huge Crater

水星表面布满了陨石坑。

The surface of Mercury is covered with craters.

这些陨石坑是小行星、流星体和彗星撞击水星表面而形成的。

These craters were made by asteroids, **meteoroids**, and comets that hit the planet's surface.

水星上还有一些极其巨大的陨石坑，被称为“撞击盆地”。

Mercury is also home to some truly giant craters called **impact basins.**

水星最大的撞击盆地直径约1 500千米。

The planet's largest impact basin is about 1,500 km wide.

它大到能完整装下整个美国得克萨斯州。

It's large enough that the entire state of Texas could fit inside!

当小的太空岩石与行星大气层相撞时，它们会解体或燃烧起来。水星的大气层非常稀薄，无法阻挡这些岩石，因此它的表面只能一次次地受到撞击！

When small rocky space objects hit a planet's atmosphere, they break apart or burn up. Mercury's very thin atmosphere cannot stop these rocks, so its surface gets hit again and again!

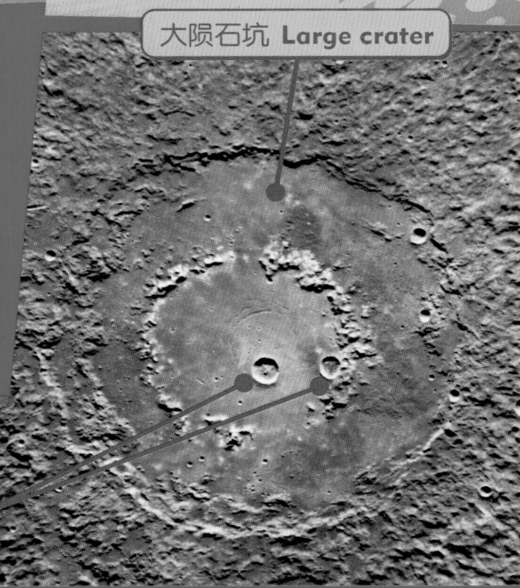

水星上有个巨大的撞击盆地，被命名为“卡洛里斯盆地”。这张水星照片里的黄色区域就是卡洛里斯盆地。

Mercury's gigantic impact basin is called the Caloris Basin (kuh-LOR-iss BAY-sin). The large yellow area on this picture of Mercury is the Caloris Basin.

小水星，大太阳！
Tiny Mercury, Giant Sun!

每隔几年，当水星从太阳前面经过时，人们便可从地球上看见它。

Every few years, Mercury can be seen from Earth as it passes in front of the Sun.

这种经过日面的运动被称为“凌日”。

This movement across the Sun is called a **transit**.

当水星经过巨大的太阳时，这颗行星看起来就像一个小点！

As Mercury moves across the giant Sun, the planet looks like a tiny dot!

人们不能直视太阳，因为这会严重伤害眼睛。

People should never look directly at the Sun because it will badly damage their eyes.

所以，科学家会使用特殊设备拍摄凌日现象。

So scientists use special equipment to photograph a transit.

这样，大家就可以安全地欣赏这一奇观了。

Then everyone can safely look at photos of this amazing event.

太阳黑子 Sunspot

这张照片摄于2006年11月，显示了水星从太阳前经过。
This photo shows Mercury passing in front of the Sun in November 2006.

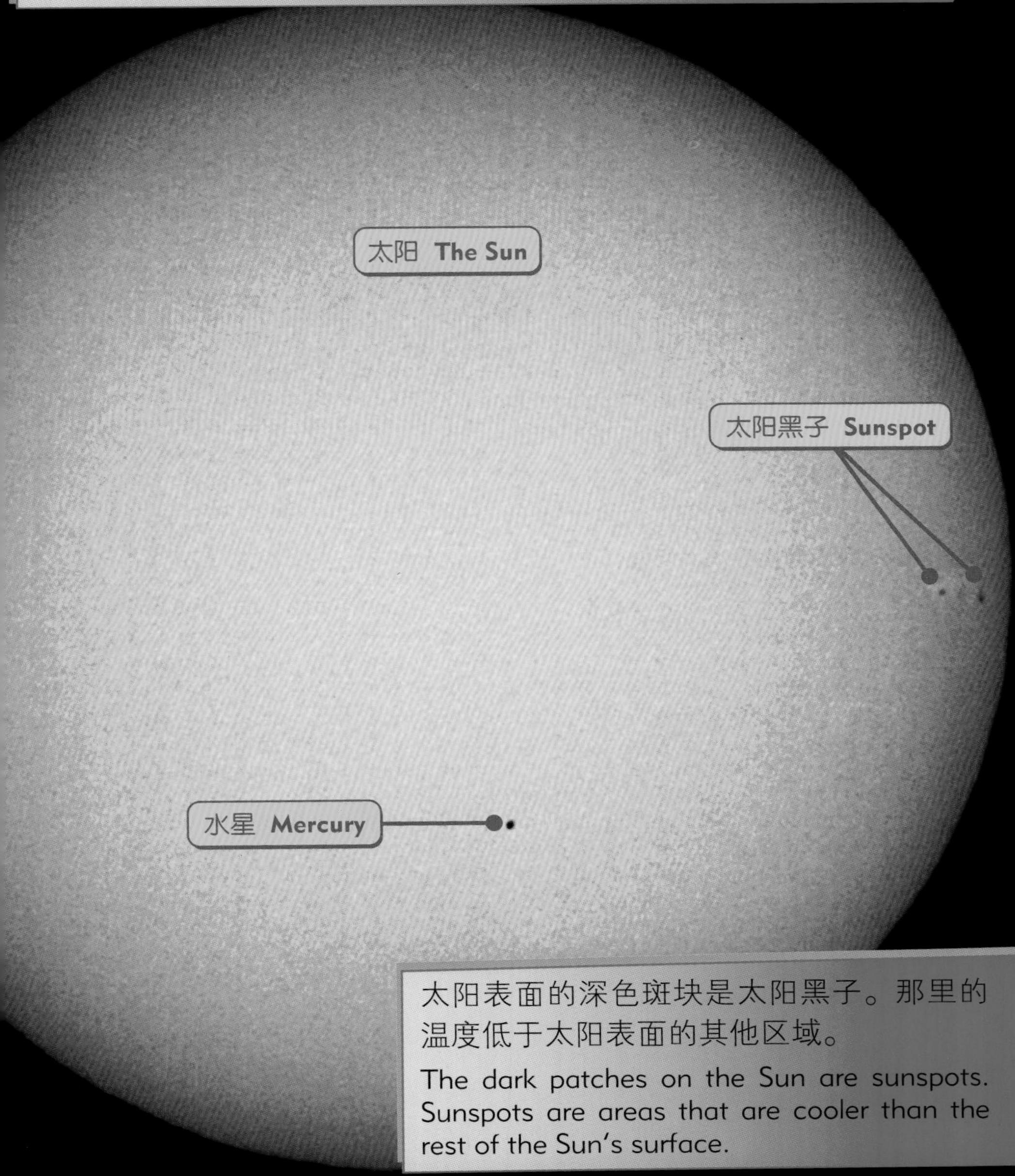

太阳表面的深色斑块是太阳黑子。那里的温度低于太阳表面的其他区域。
The dark patches on the Sun are sunspots. Sunspots are areas that are cooler than the rest of the Sun's surface.

探测水星的任务
A Mission to Mercury

2004年8月，一枚火箭从地球发射升空。

In August 2004, a rocket blasted off from Earth.

火箭上搭载了太空探测器“信使号”。

Aboard the rocket was a space **probe** called *Messenger.*

“信使号”的任务是飞往水星并开展研究。

Messenger's mission was to travel to Mercury and study the planet.

“信使号”飞行了6.5年才抵达目的地。

It took six and a half years for *Messenger* to reach its destination.

2011年3月，这个太空探测器开始环绕水星公转。

In March 2011, the space probe began orbiting Mercury.

2004年，火箭搭载着“信使号”从美国佛罗里达州卡纳维拉尔角升空。

Messenger blasts off from Cape Canaveral, Florida, in 2004, aboard a rocket.

这张照片显示了“信使号”绕水星公转时的场景。有时它离水星地表的距离只有200千米。

This picture shows how *Messenger* might have looked orbiting Mercury. Sometimes it was just 200 kilometers above the planet's surface.

惊人的发现
Amazing Discoveries

“信使号”在执行探测任务的过程中发现，极度炎热的水星竟然存在冰。

During its mission, *Messenger* discovered that there is ice on super-hot Mercury.

这些冰都集中在陨石坑的底部。

The ice is at the bottom of craters.

由于这些陨石坑实在太深了，所以太阳的光和热没法影响到底部的冰！

The craters are so deep that the Sun's heat and light never reach the ice!

“信使号”拍摄了水星上几处巨崖的照片。

Messenger also took photos of huge cliffs on the planet's surface.

2018年10月，又有一个装载着两颗人造卫星的太空探测器离开地球，前往水星。

In October 2018, a new spacecraft carrying two satellites left Earth headed for Mercury.

此次探测任务被命名为“比皮科伦坡任务”，那些人造卫星将在2025年的12月开始投入对水星的研究工作。

Named the BepiColombo mission, the satellites will begin studying Mercury in December 2025.

“信使号”拍摄的水星南极的陨石坑。

Messenger took this picture of craters at Mercury's south pole.

这张图片是计算机利用“信使号”收集的信息而生成的。它显示了水星上的一座巨大悬崖。这座悬崖高1.6千米，绵延数百千米。

This picture was created on a computer using information from *Messenger*. It shows the giant cliffs on Mercury. The cliffs can be 1.6 km high and hundreds of kilometers long.

两颗比皮科伦坡卫星将与太空探测器分离，并围绕水星公转。它们会研究水星的形成，也会搜集水星地表和水星内部的信息。

The two BepiColombo satellites will separate from the spacecraft and orbit around Mercury. They will study how the planet formed. They will also collect information about the surface and the inside of Mercury.

有趣的水星知识
Mercury Fact File

以下是一些有趣的水星知识：水星是距离太阳最近的行星。

Here are some key facts about Mercury, the closest planet to the Sun.

水星的发现
Discovery of Mercury

不用望远镜也能在天空中看见水星。人们早在古代就发现了水星。

Mercury can be seen in the sky without a telescope. People have known it was there since ancient times.

水星是如何得名的
How Mercury got its name

这颗行星是以古罗马信使之神的名字命名的。

The planet is named after Mercury, the ancient Romans' messenger of the gods.

行星的大小
Planet sizes

这张图显示了太阳系八大行星的大小对比。

This picture shows the sizes of the solar system's planets compared to each other.

太阳 Sun

水星 Mercury

金星 Venus

地球 Earth

火星 Mars

木星 Jupiter

土星 Saturn

天王星 Uranus

海王星 Neptune

水星的大小
Mercury's size

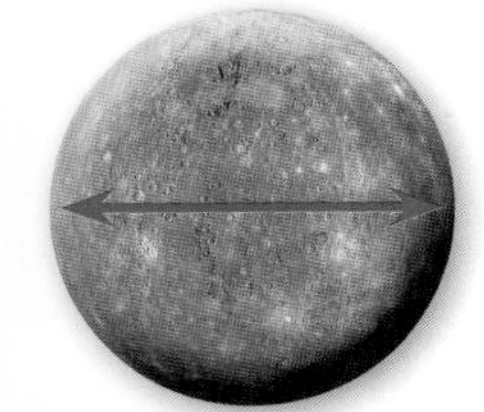

水星的直径约4 879千米

4,879 km across

水星自转一圈需要多长时间
How long it takes for Mercury to rotate once

1 408个地球时

1,408 Earth hours

水星与太阳的距离
Mercury's distance from the Sun

水星与太阳的最短距离是46 001 009千米。

水星与太阳的最远距离是69 817 445千米。

The closest Mercury gets to the Sun is 46,001,009 km.

The farthest Mercury gets from the Sun is 69,817,445 km.

水星绕太阳的轨道长度
Length of Mercury's orbit around the Sun

359 976 856千米

359,976,856 km

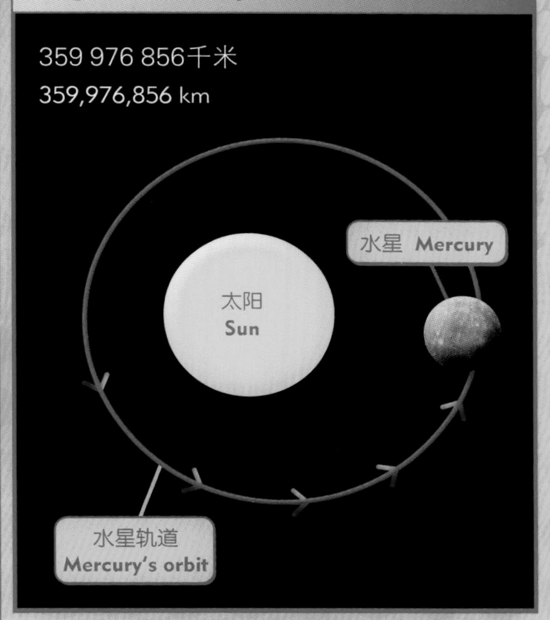

水星围绕太阳公转的平均速度
Average speed at which Mercury orbits the Sun

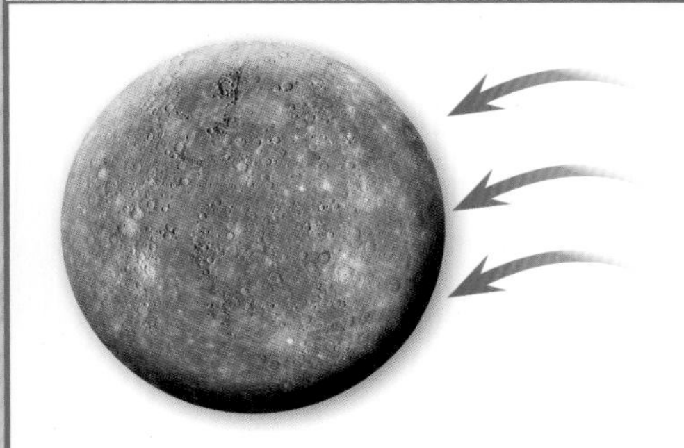

每小时170 503千米

170,503 km/h

水星上的一年
Length of a year on Mercury

88个地球天

88 Earth days

水星的卫星
Mercury's moons

水星没有卫星。

Mercury has no moons.

水星上的温度
Temperature on Mercury

最高温度：427摄氏度
最低温度：零下173摄氏度

Highest: 427°C
Lowest: −173°C

动动手吧：制造陨石坑！

Get Crafty : Crater Maker!

太空岩石撞击水星表面后形成了陨石坑。通过使用鹅卵石和熟石膏，你也可以制造水星陨石坑。

你需要：

- 一只碗
- 熟石膏
- 水
- 一只勺子
- 一个锡纸烤盘
- 报纸
- 鹅卵石或小石头

1. 将熟石膏倒进碗里。加入一些水并搅拌，直至这些混合物看起来像细腻的煎饼面糊。继续添加石膏或水，以取得更多的混合物，因为你需要足够的混合物来装满烤盘。

2. 将混合物倒入烤盘里，再将烤盘放在地板上（烤盘下垫些报纸），等待混合物成型。不时用手指触碰混合物，当你感觉混合物变得像刚融化的冰淇淋时，就可以动手制造陨石坑了。

3. 将鹅卵石咂进烤盘里。它会立即将石膏混合物砸出了一个坑！小心、迅速地取出鹅卵石，接着换块鹅卵石再砸一次。

4. 当石膏混合物布满坑的时候，将它静置，等它慢慢变硬。这大约需要30分钟。然后小心地将模型从烤盘中取出。

词汇表 Glossary

小行星 | asteroid

围绕太阳公转的大块岩石，有些小得像辆汽车，有些大得像座山。

大气层 | atmosphere

行星、卫星或恒星周围的一层气体。

彗星 | comet

由冰、岩石和尘埃组成的天体，围绕太阳公转。

陨石坑 | crater

圆形坑洞，通常由小行星和其他大型岩石天体撞击行星或卫星表面而形成。

矮行星 | dwarf planet

围绕太阳运行的圆形或近圆形天体，比八大行星小得多。

撞击盆地 | impact basin

非常大的陨石坑，直径可达数百千米。

流星体 | meteoroid

从彗星或小行星上脱落的一小块岩石或尘埃。

公转 | orbit

围绕另一个天体运行。

行星 | planet

围绕太阳公转的大型天体。一些行星，如地球，主要是由岩石组成的。其他的行星，如木星，主要是由气体和液体组成的。

探测器 | probe

不载人太空飞船。通常被送往行星或其他天体，用于拍摄照片并收集信息，由地球上的科学家操作控制。

自转 | rotate

物体自行旋转的运动。

太阳系 | solar system

太阳和围绕太阳公转的所有天体，如行星及其卫星、小行星和彗星。

凌日 | transit

小型天体从大型天体表面经过。例如，水星经过太阳表面就是水星凌日。